Tic, Tic, Tic, Tic, Tic

Tic, Tic, Tic, Tic, Tic

Joe Adcock

ISBN: 1544127359

ISBN 13: 9781544127354

Library of Congress Control Number: 2017903157

CreateSpace Independent Publishing Platform

North Charleston, South Carolina

Dedicated to *you*, the reader.

For I tell you, that many prophets and kings have desired to
see the things which ye see, and have not seen them; and to
hear the things which ye hear, and have not heard them.
—Luke 10:24

Bible verses in this text use Noah Webster's translation of the Bible. Published in 1833 as the *Common Version*, Webster used the King James Version as a base, updating the grammar and lexicon. Webster, one of the great American educators of the nineteenth century, said that education was "useless without the Bible." This text, published prior to 1923, is in the public domain in the United States.

Today, as I'm writing this sentence, the date is Monday, August 15, 2016, forty-two years since I first stood upon Seal Beach Pier on June 8, 1974, and thought about the Star in the East.

∞*– –*– –*– –|(*)– +((*))…

Printouts

Star in the East at Jerusalem
Solar eclipse at Jerusalem
Star in the East, view from space
Future star, view from space
Future star at Jerusalem
Ten Commandments star at Jerusalem
Ten Commandments star, view from space
Evening star at Jerusalem
Evening star, view from space
Adam and Eve star at Jerusalem
Adam and Eve star, view from space
Future-Future star at Jerusalem
Future-Future star, view from space

Drawings

Timeline drawing
Early program drawing of Star in the East at Jerusalem
Early program drawing of Future star at Jerusalem
Early program drawing of Ten Commandments star at Jerusalem
Early program drawing of Adam and Eve star at Jerusalem

Preface

My dad lived to be eighty-five years old. Sitting at his kitchen table when he was quite old, I showed him what was in this book. Dad said dismissively, "That's nice, son." The facts about stars and biblical events didn't connect. I sat there, unable to communicate or convey the meaning of the stars. I couldn't break through to open his eyes, ears, mind, or heart. Before he died, he told my brother, "I'm afraid."

I was powerless then, and I am powerless now. So I ask *you* to open your eyes, your ears, your mind, and your heart to the *stars*. The time is ticking…

Joe

Acknowledgments

T hanks to Michael at the corporate site of www.simulationcurriculum.
com for allowing the use of its astronomy printouts. Those printouts
made real the statement that a picture is worth a thousand words. I
used an astronomy program called "Starry Night Backyard" in my search-
ing/research, found at www.starrynight.com.

Thanks to Ms. M for my first astronomy program at Christmas.

Thanks to writer/editor Ms. Jody Duncan for her very timely assistance. Her insights on writing helped me stay focused on continuity and
not to expect the reader to read my mind.

Thanks to Pastor Scott Elgersma for adding what I call polish to the
manuscript. I tried to address every suggestion/correction indicated, including several rewrites.

I would like to thank my wife, Barbara, for spelling, word assistance,
and encouragement and my family for support when needed.

To CreateSpace, I like to give credit where credit is due. High marks
to you and five real stars. Special thanks to the editors: Karen, Jennifer,
and Deedee.

Introduction

Y^{*ou…*} Where are *you* in time? Do you know? I know. I am possibly the only one who knows, but only because I discovered a *star*, a benchmark in time, with an astronomy program. The star is part of a timepiece that is ticking at this very minute, even as you are reading this. The ticking is leading us toward our future destiny. The timepiece has almost equal periods of time that most of you will recognize easily. This book is to *you*, for *you* to find out about *you*.

This concerns *you* and events that coincide with special types of stars. First, the stars I am referring to are not stars you normally see appearing in the night sky or the romantic stars of which poets write. The stars in this book are three or four planets converging together as a visible group and dispersing, only to appear periodically at recurring intervals of time. Although technically incorrect, the term "star," coined ages ago, describes a planetary convergence. The singular known "star" that the term belonged with is the Star in the East associated with the birth of Jesus. There are other events connected to other stars.

It may surprise you that everyone generally knows about the events, but people hardly know anything about the stars. More unusual is that these stars are still in the process of occurring. We are, in fact, between two stars right now. The real story to learn involves the events that coincide with the stars. You will encounter a thought-provoking specialty subject, and *you* are involved! I know the past, and that is no big deal. I know the future, and that is a big deal, but only because of the stars/planetary convergences. Astronomy programs have computerized the data that will allow you to travel to almost any place and time in the solar system. This book's primary focus is on planetary orbits in the past, present, and future. Views of the sky and views from space led to discoveries in the past and in the future and more knowledge afterward. I am profoundly grateful to the corporate site of www.simulationcurriculum.com for allowing me to use printouts from its astronomy program called "Starry Night Backyard," www.starrynight.com.

As an example of the program's versatility, later in the text, there is a reference to a planetary alignment that I believe helped to part the Red Sea. What did not make it into the text is that I wanted to see if an obvious relationship with the moon caused a low tide contributing to the parting of the Red Sea, and the astronomy program allowed me to view Earth from the moon. That view was not helpful, but I still believe the moon helped to part the Red Sea.

Many religions and many holy books exist in the world, but the Christian Bible has the only information, as far as I know, about ancient and future events connected to stars.

I will do my best to provide interesting background information about significant events that brought this book into your hands. On our journey together, we will discover a key to the stars found in an astronomy computer program. *You* are the reason for this journey. Once you and I use the key, we will unlock a timeline. Our trips will take us from the present to the future; then we will go to the past and then revisit the future and a future-future mystery.

A type of code using unusual symbols throughout the text indicates your position in time. The code is simply a condensed version of a timeline. It is about five crucial morning stars represented by an asterisk (e.g., *). Three morning stars are already in the past, two stars are in the future, and one of those poses a mystery. Time spans of approximately 1,125 years will be represented by an en dash (e.g., –), and the last time span has 1,125 years plus an additional 495 years, symbolized by a plus sign (e.g., +). An asterisk with parentheses represents a future star (*). An asterisk within double parentheses represents a future-future star ((*)). Infinity (e.g., ∞) is the time before the first star, and ellipsis (e.g., …) is the time after the last star. A vertical line (e.g., |) is *you*, your position in time. A sixth star, an evening star in the past, is not included in the timeline.

The code acts as visual shorthand for the timeline, and it begins in the text with a single asterisk * to represent the Star in the East and builds to coincide with the information. A detailed master timeline drawing is included later in the text with a lot more significance. The completed summarized code follows.

∞*– –*– –*– –|(*)– +((*))…

Onomatopoeia is a word that describes itself by its sound, such as "meow" or "bang." I feel that the sounds of "Tic, Tic, Tic, Tic, Tic" in a row communicate a common understanding of marks in time. I am comfortable using five "Tics" for the title. "Tick" can be a sound of a clock or a parasite that feeds on blood or a tick-off mark. Although "tic" is a facial twitch, it does not relate to the sound of the word. Ticking is in the text as a sound of a clock passing time.

There is some information that I was not able to provide, and that is the brightness of the stars. I tried it and decided to leave it to the professionals.

Also left for the professionals is the question: Did the tides part the Red Sea during the Exodus from Egypt, with the help of a planet alignment?

These answers are obtainable—ask Watson!

CONTENTS

CHAPTER 1   THE STAR OF THE EAST , OF BETHLEHEM, OF JESUS
             CELESTIAL NAVIGATION
             ASTRONOMY PROGRAM

CHAPTER 2   QUOTATION MARKS
             FUTURE STAR
             AH-HA

CHAPTER 3   STARS PAST - PAST
             LAW
             CONTRACT

CHAPTER 4   MYSTERY STAR
             TIMELINE
             DAYS OF GOD
             MAGNETIC REVERSAL

CHAPTER 1

THE STAR: OF THE EAST, BETHLEHEM AND JESUS CELESTIAL NAVIGATION ASTRONOMY PROGRAM

*Y*ou...

Hi, my name is Joe, and I am going to be your guide to some very special stars and other information. We will be traveling to times that co-incide with events that you may have some knowledge of; only now there are new facts and insights for you. This journey will be through thousands of years in time. Soon, to begin our journey, we are going to Jerusalem. Not every event happens in Jerusalem, but we can observe from there. An astronomy program will provide our transportation, hand in hand with an-cient writings in the Holy Bible. The Starry Night Backyard Corporation has granted permission to use printouts from its astronomy program. As a visual aid, you will see that a printout is worth a thousand words. I am the first (as far as I am aware) and, as of now, the only one who has learned how to navigate to events in the past as well as the future. It may have been due to some luck in finding the Star in the East or perhaps an educated hunch. I have many interesting and exciting things for you to see concern-ing the past and the future. Let's check out the breakthrough and see what I discovered.

A star in ancient times has presented a mystery throughout the ages. What was the star? How did the star appear, and what did it look like? We are sure the star appeared because, recorded in history, it marked the birth of Jesus. You will learn very soon that this star concerns you and that this star is part of something bigger—much bigger—than you can imagine; it unlocks the key to time—a benchmark.

This benchmark involves you more than you think, and it is gratifying to have you on the journey. I don't know much about you, and I certainly don't know your name, what you look like, or even if you're male or female. So do I know anything regarding you? Yes, I do. I know where you fit in time and a little of why. Until now, I have been the only one to know. So, how did I come to get this information? It is my intention to show you soon.

First, you need a little background on how and why I obtained the information that has found its way to you. The best way to explain is by going back to a series of what seem like unrelated activities. So, let's connect the dots.

First dot: If I had not met a special sailor on the job, then I never would have found the star.

Early in my career, I worked at a job where the foreman talked about sailboats every day. He didn't even have a sailboat, to my knowledge. What I knew about boating, any boating, was very little. Every day, day after day, I listened to talk of sailboats, sailboats, and more sailboats. He told of cruising to Mexico on a large sailboat as a cook. While on that trip, he trained the owners' dog to use a newspaper for its bathroom. He told another story about how he and the owners woke up to thumping noises on the hull from a school of fish surrounding the boat. During the America's Cup in San Diego, California, he helped on the towboat. He was one of my best foremen ever; I loved having him as my foreman. We had no boring break times on that job, as he told story after story after story.

Second dot: Buying a sailboat. If I hadn't bought a sailboat, I never would have found the star.

So the next most significant event started when I went to my wife and said, "Could we go look at boats?"

"Oh yeah," she said. "A powerboat."

"Ah, no, I'm thinking of a sailboat."

We ended up at a sailboat dealership and directed to a salesman. I asked, "Can you tell us about sailboats?"

As he began speaking, I noticed he looked at something distant. Then unexpectedly he said, "Can you wait here a minute, please?"

"Sure," I said, puzzled.

Shortly he came back and said, "Come on; let's go. I've got permission from new owners, and we can take their boat out." The boat was in a slip near the dealership with the mainsail up, and the salesman asked us to step onboard. He cast us off and pushed the boat into the channel; we glided silently, effortlessly, as we progressed up the channel, and we had a serene experience. The salesman didn't say a word; he didn't need to—the boat said it for him. My wife had the expression of "This is fantastic," and my return look expressed the same sentiment. Could he teach us to sail? "Absolutely," he said. "We'll go out with you until you get the hang of it." Suffice it to say, operating a sailboat has a learning curve and a language unto itself. My life is full of sailing stories, but as your guide, I need to get us back on course.

Third dot: Joining a yacht club. If no yacht club, then I never would have found the star.

We joined a yacht club and learned from sailors who had been all over the world, and they energized my travel bug. The yacht club also provided many activities, including racing and cruising, and we formed many friendships sailing. The mantra in the yacht club was "Do it now before it is too late," or before you are old and decrepit. There was another saying in the yacht club, something to the effect that "The only sure thing is change." In order to go all over the world at that time, you needed celestial navigation—a method to locate your position at sea using the sun, moon, planets, and stars.

Fourth dot: Learn celestial navigation. If no celestial navigation, then I never would have found the star.

For me, one particular milestone stands out: the day I passed my final exam in celestial navigation. I don't think my feet touched the ground for a week, I was floating so high. The knowledge that, except for the polar regions, I could untie my boat and push off to any place on the earth gave me a fantastic freedom I cannot explain. That freedom has stayed with me ever since. I've never used it for any great ocean crossing, but I know it's always there.

Celestial navigation and astronomy have many of the same descriptive terms that belong to both studies. After completing navigation classes, I took astronomy in college, and the first night of the class, the instructor put things into perspective. He said something like this: "This subject will probably never make you a dime, but it is interesting at cocktail parties." I for one appreciated that insight. If the idea of your body consisting of stardust appeals to you, then you might like astronomy also. There are things in astronomy that you might find interesting.

Regarding the lectures in celestial-navigation class, there may be some useful information for you, such as how a celestial object near the horizon can appear up to four times larger. The cause of a harvest moon at the horizon depends on air molecules to refract light traveling long distances through the atmosphere.

Apparent magnitude is the logarithmic measure of an object's brightness as it appears in the night sky. The scale looks reversed because the brighter the object appears, the smaller the number. The sun has a magnitude of (–27), the full moon (–13), and the visible planets to the naked eye vary:

–30	–25	–20	–15	–10	–5	0	+5	+10	+15	+20	+25

–27 Sun   –4.89 Venus

–13 full moon   +6 faintest stars

–2.95 Mars

–2.70 Jupiter

–0.49 Saturn

+2.45 Mercury

- For Venus a maximum of (–4.89) to a minimum of (–3.82)
- For Saturn a maximum of (–0.49) to a minimum of (+1.47)
- For Mars a maximum of (–2.95) to a minimum of (+1.84)
- For Jupiter a maximum of (–2.70) to a minimum of (+1.61)
- For Mercury a maximum of (+2.45) to a minimum of (+5.73)
- The faintest stars seen with the naked eye are (+6)

As your guide: let's get back on course.

Navigators use fifty-seven stars, the sun, the moon, and four planets for traveling the world's oceans. The four navigational planets are Venus, Mars, Jupiter, and Saturn. Mercury makes the fifth planet visible to the naked eye.

Try to imagine standing with our celestial-navigation class at the end of the Seal Beach Pier in California. It's twilight, and the cold, damp sea air has moistened everything it encounters, even me. The smell of old fish bait and fish parts wafts in the air pungently, and there are the incessant rhythms of ocean swells splashing against pilings. Seagulls squawk to other seagulls as if to say, "I found food! I found food!" It is tranquil. Our instructor already knows what we should see in the sky as student navigators. He wants us to practice a planetary sight to confirm our location. It is June 8, 1974. Mars is in the west. We have learned from the class that planets can appear to wander, or retrograde.

Having completed the planetary sight—and by then our class had done many, many sights for classroom practice—it clicks for me, an epiphany: the planets have orbits in a relatively tight range of elevations, and they retrograde. The Christmas star! A star appeared in the east when Jesus was born in Bethlehem. Could the Star in the East be the result of planets converging? That thought stayed with me for years and years. In the New Testament of the Christian Bible, Matthew contains a lone verse referring to a star. Wise men came: "Saying, Where is he that is born king of the Jews? for we have seen his star in the east, and come to worship him" (Matt. 2:2).

Fifth dot: Matthew 2:2. If no Matthew 2:2, then I never would have found the star.

Sixth dot: A hand-me-down computer. If no computer, then I never would have found the star.

My son was in college studying chemistry, and his roommate was studying computers, seemingly at the time computers were developing exponentially in speed and memory. My son picked up the information and started building computers for himself as soon as new parts became available. His continual upgrades resulted in my computer being a hand-me-down, though state of the art for that time, and only one notch below his own. The great thing, though: I had my own tech.

Now let's fast-forward: divorced, relocated to live on the boat, remarried, and now let's move on to the next significant events.

The following episode happened in the summer of 1995, and it seems very unusual. Hindsight makes this encounter even more bizarre, although at the time, I thought very little about it.

I got off work and went to the local supermarket in my dirty work clothes, and I was shopping near the end of the bread aisle in an open area across from the vegetable tables. Unexpectedly, a short, elderly woman was standing to my left and a little behind me. She handed me a white plastic spoon, a stubby pencil with no eraser, and a white sheet of paper. Her appearance was a most striking sight; she wore clothing that looked as if she had just stepped off the Oregon Trail: a floor-length dress from a hundred years ago, bonnet, and all. She said, "You have been chosen by God to be a prophet."

Then a man behind the meat counter said, "Sir, is that woman bothering you?"

I turned toward him and said, "No," and immediately I turned around to look at the woman, who had vanished. She had disappeared in an instant, with no apparent place to go without me seeing her. I looked back toward the butcher, a tall man, and he had also disappeared. I stood there alone, holding the spoon, pencil, and paper, and thinking, *What just happened?* It might have taken a hundred times longer to write this paragraph than the episode itself, which took only seconds. I finished shopping and thought, *She was just an old woman*. I took the items back to the boat and studied the paper for a short while. The paper was neat and covered in large, circular writings about a half inch across, as if someone had been practicing penmanship. There were no borders, headings, or punctuation, and the text didn't appear to be printed. It made no sense, and I threw the items away. *Just an old woman*, I said to myself, *nothing to be concerned about.*

Whether or not I am a prophet, I do not know; the only one to say that was the woman. I do not feel like a prophet, although an incident happened where Jesus said to the disciple John, "Forbid him not: for he that is not against us, is for us" (Luke 9:50).

For Christmas in 1995, I received an astronomy program as a gift from my stepdaughter, M, and that old thought came back. It had been six months since that summertime encounter, and it had been twenty-one years since my thoughts on June 8, 1974, at the Seal Beach Pier about the star of Jesus's birth.

Seventh dot: An astronomy program. If no astronomy program, then I never would have found the star. The astronomy program became the last remaining dot to connect for the discovery of the star—and it was the most crucial.

As you can tell, many dots had to line up over a very long time to get this information for you.

I am going to describe for you what it felt like when I made the discovery of the Star in the East. I wrote this quite a while ago when it was fresh in my mind; although I was by myself, I am sure I said "*Bingo*! There it is," out loud.

How the star discovery happened agrees with the subsequent description, but later facts from an updated astronomy program follow because there are discrepancies in the original program with the date, as well as the appearance of a fourth planet Mercury.

Imagine that you are with me while I am sitting on a two-foot ladder at the keyboard with a cushion to sit on made for throwing.

Soon after Christmas in 1995, one evening after dinner, that old thought about the Star of the East came back. The boat harbor was quiet, and a warm glow of light cast about the cabin, highlighting the oiled teak. It being a relaxing time of the day, I decided it seemed like a good time to find out what the astronomy program could do. There were no expectations for me; I turned the computer on and started the astronomy program. I set my location at Jerusalem, Israel, because it should be the nearest city to Bethlehem on the list of cities, and adjusted the program so I could look eastward. Beginning with January in year AD 1, I clicked the mouse to move forward one day at a time, following day by day while watching the planet orbits. In the year AD 1, October 22, Venus, Mars, and Jupiter began to converge early in the morning. "*Bingo!* There it is"—the Star in the East! I got excited; I discovered the Star concealed in an astronomy program. Was I the first person to see it in two thousand years? I believe so. The star didn't take very long to find, and I remember thinking that anyone could find the star if interested. After closing the program, I went to sleep late that night, excited to have found the star. The star would turn out to be a key and a benchmark in history.

The first astronomy program had differences from the one I enjoy now; that first program is out of date and inaccurate. So that you can see what I did in the beginning, I have included, at the back of the book, four depictions from that first program. Not shown in the depictions are the orbit lines, which I used to locate the star. They had a different color for each planet and appeared to go vertically straight up from Jerusalem. Because the program is so old, I am unable to locate or obtain the copyright permission; thus, I made depictions for you.

October 22 is *not* the correct date, and that early astronomy program did not include a fourth planet, Mercury. The website www.starrynight. com has provided accurate astronomy printouts for you to see. Included in the main text are the *correct* views only.

Perhaps you might be wondering how this information should be a concern to you. Do *you* know where *you* are in time? I do. Celestial-navigation class was the trigger that opened the idea that planets formed the Star in the East.

As your guide, I would like you to look at the Star in the East; the "Starry Night Backyard" program has an accurate date for the "star" or planetary convergence: November 3 to 7 in the year AD 1, just before dawn, around 5:00 a.m. When I first used this program, the planet Mercury appeared, and that was another big astonishment to me. Unexpected is an understatement. Four of the five visible planets together certainly did make a star. The "star" written about in the ancient text demonstrates a lack of knowledge regarding a planetary convergence. In a convergence, the planets need to stay in close proximity before spreading too far away from each other to be a "star," which usually takes about three days. Sunday, November 6, is my chosen date, because that seemed to be the best view of the convergence, but the birth of Jesus could be the day before or the day after.

*

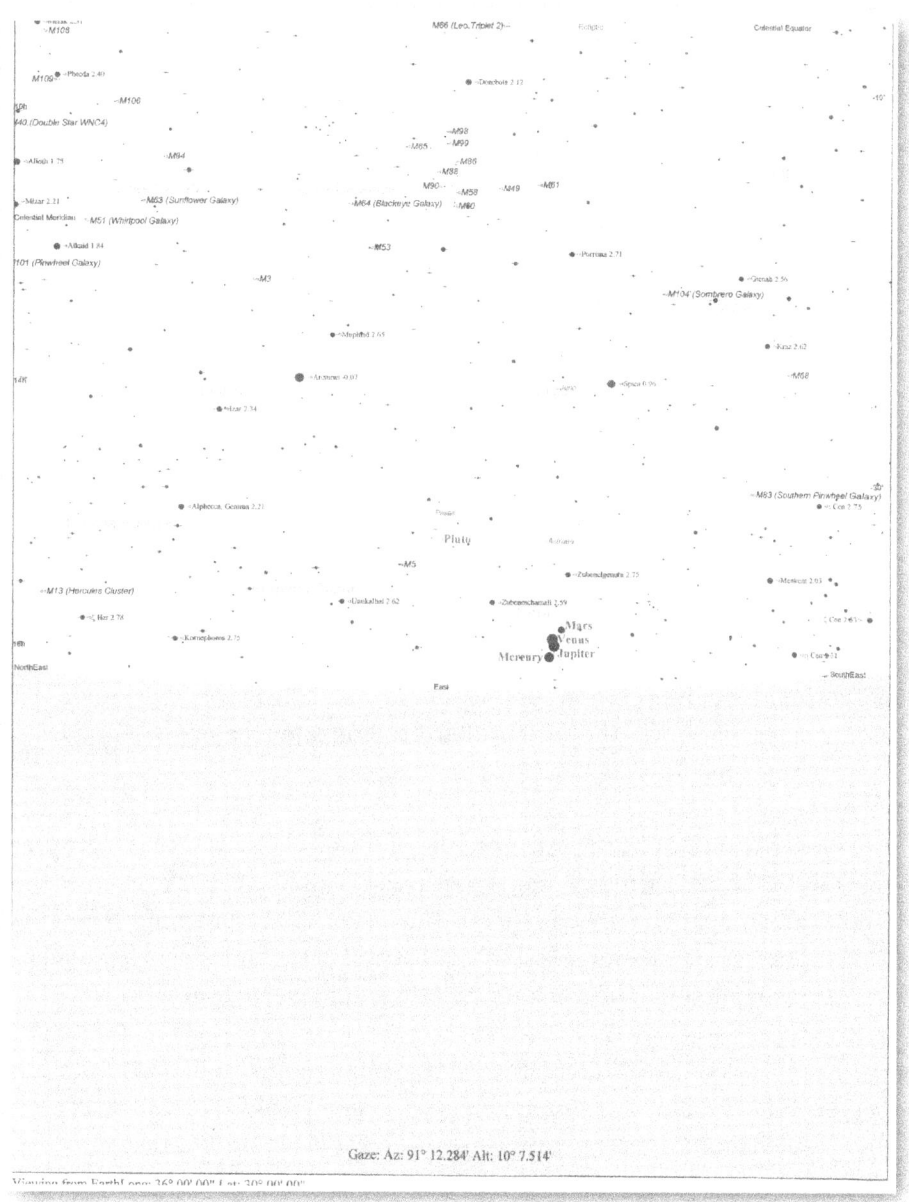

Gaze: Az: 91° 12.284' Alt: 10° 7.514'

Viewing from Earth | app: 36° 00' 00" Lat: 30° 00' 00"

Star in the East, at Jerusalem, November 6, AD 1, at 5:00 a.m.
Image courtesy of Starry Night Astronomy™.

If you want to see the star on your own astronomy program, the country of Israel has "daylight savings time," and that feature may be automatically set in the program. The time difference might mean that the star is below the horizon, where you cannot see it. It would be best to uncheck the "daylight savings" box. Otherwise, you might be one hour off and may need to change to an earlier time to see the star. If you still do not see the star, look behind the trees by changing the hour or set for a flat horizon.

The fact that I was able to discover the star kept within a computer program means more than a found star. It provides an absolute *proof* that Jesus was a real person, especially considering the historical evidence recorded in Matthew 2:2 about the star. It is remarkable that we know within a day or so when Jesus was born. Think of it—you are in the past seeing "the Star in the East" at the birth of Jesus. Are you as excited as I was? Something else happened when the star appeared, because of Jesus. His birth affects you and me in a direct and personal way. How? *Grace!* He ushered in our age of grace, and we are still living in that age. Grace is unmerited love, mercy, and kindness from God. Now, as your guide, it is with great satisfaction that I have delivered on some of my promise to *you* that I wrote this for you.

As to the reference to wise men, after I found the star, I set the program to move a day at a time. When I made the planets go back in time, they would move apart. If I went forward in time, the same thing happened, they moved apart. So, I am assuming the wise men, by tracking the planet orbits, concluded that at some point they would come together.

You just saw the Star in the East. What might you be thinking to yourself? *How is that a consequence to me?* Conceivably, it is a *proof, evidence* that Jesus becomes real and not just some words in an old text.

Let this soak in a bit: if you were there in person two thousand years ago to see the star, would you have felt in a different way than you do now? My question to you inquirers: Isn't a monitor or printout of the star similar to being there? Does the star become real to you? You are, in essence, there. What will you do now that you have seen the star?

Feel the experience for a moment and imagine yourself there about two thousand years ago, looking at a bright, shining star; before you is the little town of Bethlehem. It is autumn, and a very clear, cold breeze gusts and tosses your hair back, and the wind blowing in your face makes your eyes water. The breeze carries a waft of smoky aroma from the town and the faint sound of an unseen approaching caravan that seems a great distance away. It is a little before sunrise, and the subdued light is magic this time in the morning. You know that somewhere in town, a baby has been born that will change everything. In nearby fields shepherds remain watching over their flocks, an angel of the Lord appears…Did you feel the sensation of being there? The rest of the story happens in the second chapter of Luke…

At this point, the idea that I had a key never occurred to me. It would turn out to be the first key of four. This star exists as part of something much bigger than I may have been aware of at the time.

I have had several astronomy programs, and each program had some of the stars included. Oh yes, you will learn about more stars. Certain programs do not have the range needed to see all the stars. Each program has differences in features—some better than others. One I gave away had a numbering-system feature, and the original disappeared at my kids' house.

CHAPTER 2

QUOTATION MARKS
FUTURE STAR
AH-HA

Who is Jesus? He was the person at whose birth a star appeared. Most of you have heard of Jesus, but I feel the need to highlight one of many scripture verses about him. He came for *you*. This is about *you*. His mission in this verse focuses on *you*. The emphasis of his mission remains ongoing, and even now, his mission continues: "In this was manifested the love of God toward us, because that God sent his only begotten Son into the world, that we might live through him" (1 John 4:9).

Jesus lived within "celestial quotation marks," the birth star and the day of his crucifixion. The Bible records the crucifixion of Jesus in Matthew 27:45: "Now from the sixth hour [noon] there was darkness over all the land to the ninth hour" [3:00 p.m., midafternoon]. In ancient Israel, the day ended at the twelfth hour, sundown, which is our 6:00 p.m. There was an eclipse of the sun by the moon on Friday, March 19, AD 33, from noon to 3:00 p.m., and then the sun came out.

He died Friday, the ninth hour (3:00 p.m.), as the sun came out, the exact time to slay the lambs for Passover. Friday was the Preparation Day for the Passover for the Jewish people and Good Friday to Christians. The Passover commemorates the last of ten plagues in Egypt to free the Jewish people from slavery. The angel of death "passed over" them if the shed blood of a lamb without blemish was put on the entrance to their homes. How symbolic—the shed blood of Jesus who is without sin is the Passover that delivers us from the angel of death, and it is the entrance to the door if you accept his grace. That same Friday fell on the first full moon, the start of the spring equinox. How symbolic—spring celebrates new life.

The next day, Saturday (Sabbath), arrives as a day of rest for the Jewish culture. Saturday is also the first day of the Festival of Unleavened Bread, and unleavened signifies "without sin." How symbolic—the death of Jesus, who was without sin, atoned for our sins.

The following day, Sunday, March 21, AD 33, the first day of the week, he arose from the dead. That day began the counting of the Omer, forty-nine days or seven weeks commemorating the time the Israelites left Egypt until they received the Ten Commandments. An Omer consists of a sheaf of barley, the first grain crop to ripen, the beginning of the harvest, the first fruits of the harvest. How symbolic—scripture says Jesus became the first fruits of them that slept. For Christians, the day of Pentecost corresponds to fifty days after Jesus died and the arrival of the Holy Spirit.

Only the God of the universe with celestial foreknowledge can set up a fantastic arrangement this complicated.

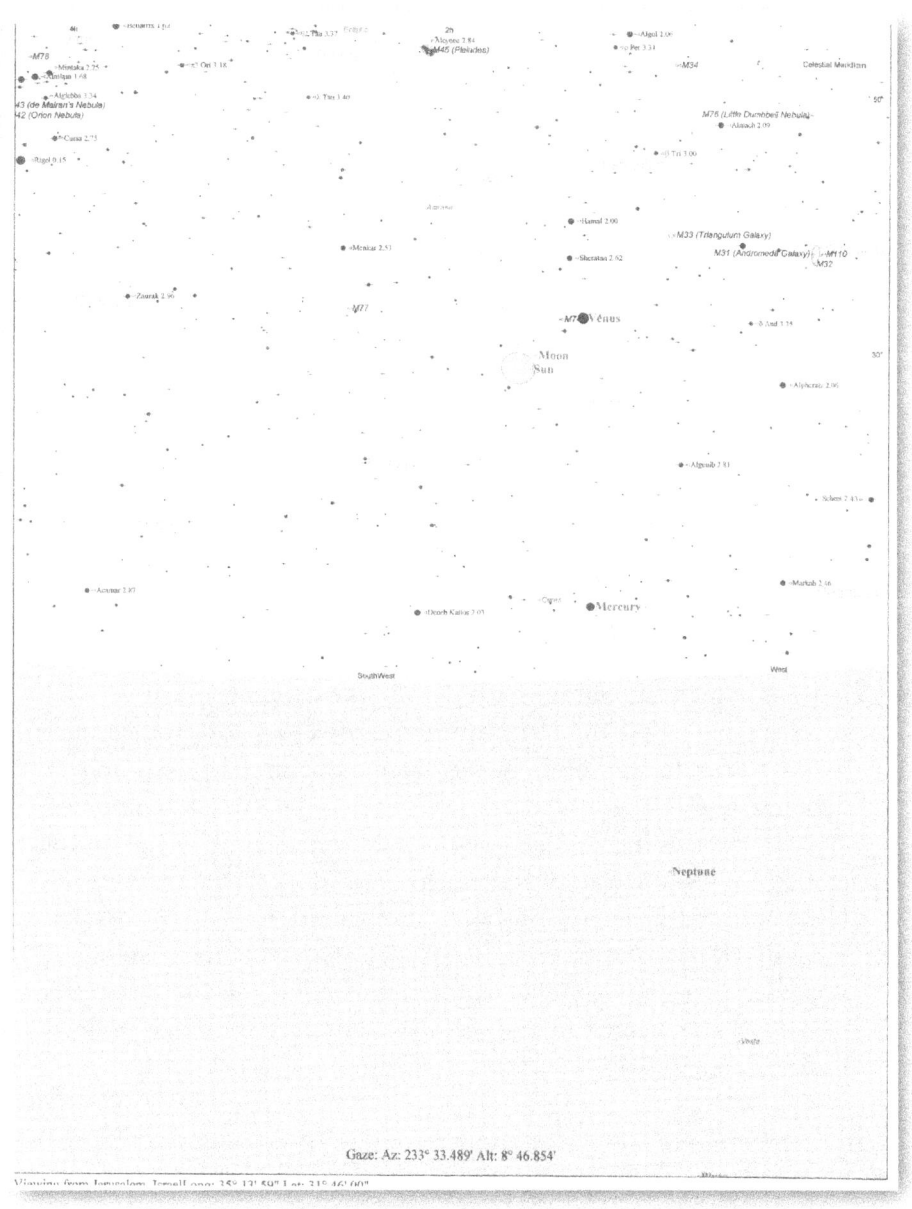

Solar eclipse at Jerusalem, Friday, March 19, AD 33, at 3:00 p.m.
Image courtesy of Starry Night Astronomy™.

Having provided evidence of the birth and death of Jesus with the astronomy program, the fact that it corroborates scripture seems amazing. As I said before, this is about *you*. What, then, is the big picture? Jesus came to restore, for *you*, the loss caused by sin from the time of Adam and Eve.

A proof of Jesus might be meaningless to many of you, because after Jesus rose from the dead, he said to his disciple Thomas, "Thomas, because thou hast seen me, thou hast believed: blessed are they that have not seen, and yet have believed" (John 20:29).

In the following scripture, the resurrected Jesus appears to be speaking to his apostles. This is the account of what he said before his departure. After speaking, Jesus went into heaven. Moreover, he has been gone ever since. Two men later in the verse relate that the way he left is the same way he is coming back again.

> And he said to them, It is not for you to know the times or the seasons, which the Father hath put in his own power. But ye shall receive power after the Holy Spirit is come upon you: and ye shall be witnesses to me, both in Jerusalem, and in all Judea, and in Samaria, and to the uttermost part of the earth. And when he had spoken these things, while they beheld, he was taken up; and a cloud received him out of their sight. And while they looked steadfastly towards heaven as he went up, behold, two men stood by them in white apparel; Who also said, Ye men of Galilee, why stand ye gazing up to heaven? this same Jesus who is taken from you into heaven, shall so come in like manner as ye have seen him go into heaven. (Acts 1:7–11)

Jesus said, "It is not for *you* to know the times or the seasons." Who then are the "you" he references? I believe he was addressing the apostles. Soon *you* shall see that the times and seasons are not hidden from *you*. I can prove with the astronomy program within a day or so when Jesus is coming back in the future. He is coming back for *you*.

I want you to know how and why the Bethlehem star of Jesus is the key to time. The astronomy program offers two perspectives: one view from outer space, showing the whole solar system, and the other view from space showing the inner solar system. The innermost solar system displays orbits and locations of planets closer to the sun, the planets we can see without a telescope. With the innermost solar system, I wondered what the Star in the East would look like viewed from that perspective. It happened to be dramatic: four planets in a row, the sun, and then the earth. This alignment of the planets reflects the sunlight, so they are visible from Earth. In this view, I had the second key, the *inner solar system* viewpoint—only I didn't know it was a second key. Even unknown to me, I unaware was about ready to turn the key. It is not exactly the kind of key to unlock doors; it is a key to unlock time. I found the key by accident, and I am excited to share with you.

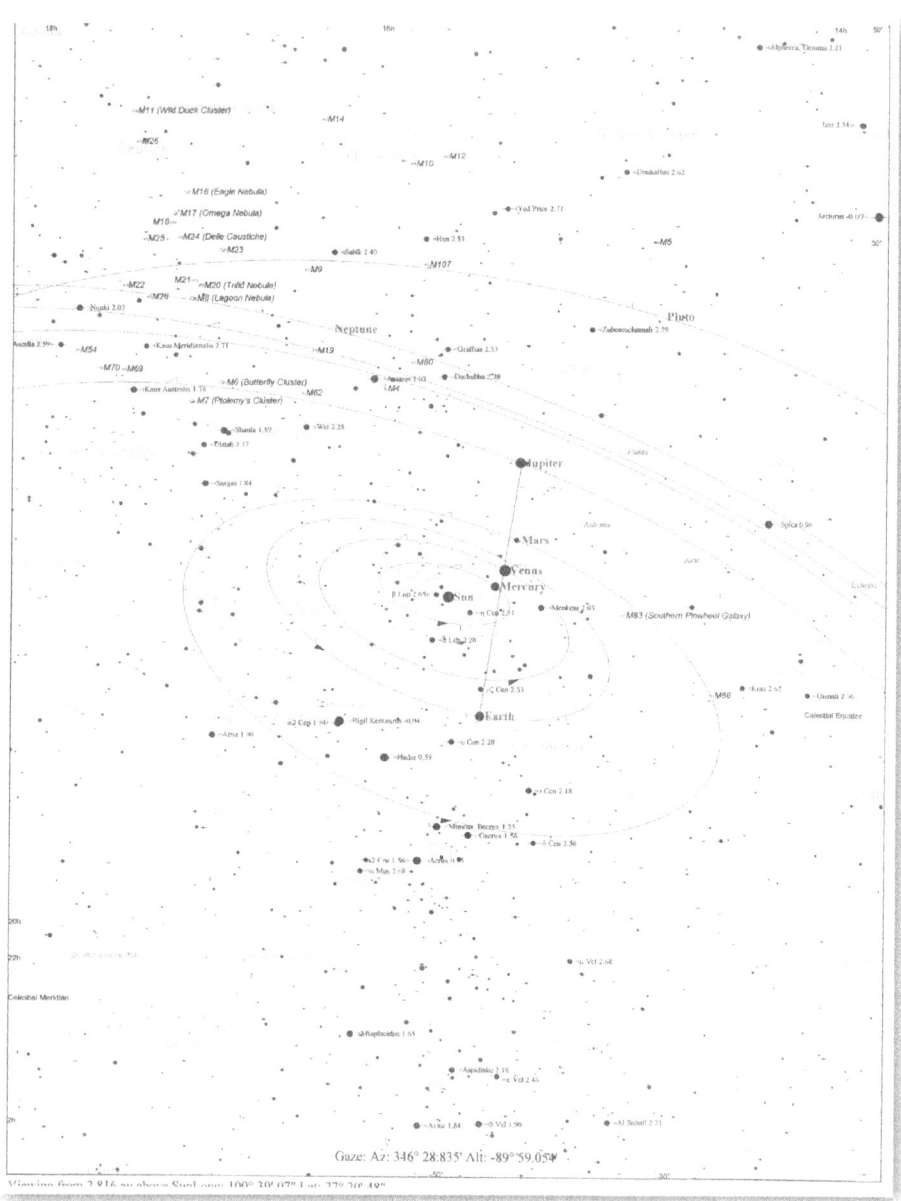

Star in the East, view from space. Image courtesy of Starry Night Astronomy™.

As your guide, please take a breath, because we are going on a breathtaking journey into the future. I have already been on the journey and have come back to act as your guide. Why the idea occurred to me, I don't know, but perhaps a month or two after I had discovered the Star in the east, I decided to check the future for a star. My thinking was, why bother to look into the past beyond the Star in the East? I knew the past, and there were no stars, to my knowledge. At this point, I must explain a useful astronomy-program feature, which was the ability to slow the rotation of planets down to a crawl; it gave me time to focus. We have the ability to recognize patterns easily, and that slow movement of the planets around the Sun was a feature—so valuable that I am sure that without it, I would not have been able to recognize patterns. A similar pattern would turn out to be my third key for new discoveries, and with it I am going time-traveling again, only this time it will be into the future utilizing three keys: first the *star*, second the *inner-solar system viewpoint*, and third *pattern recognition*. Sitting at the keyboard and using the inner-space point of view, I went looking in the future for a pattern similar to the Star in the East. After searching 256 years of time, a comparable pattern occurred. It was striking to see the same pattern come into view. Finding the pattern was unexpected, and my surprise is an understatement! Still, from that inner-space viewpoint, I could not be sure it would appear as a star when seen from Jerusalem. Would it be a future star?

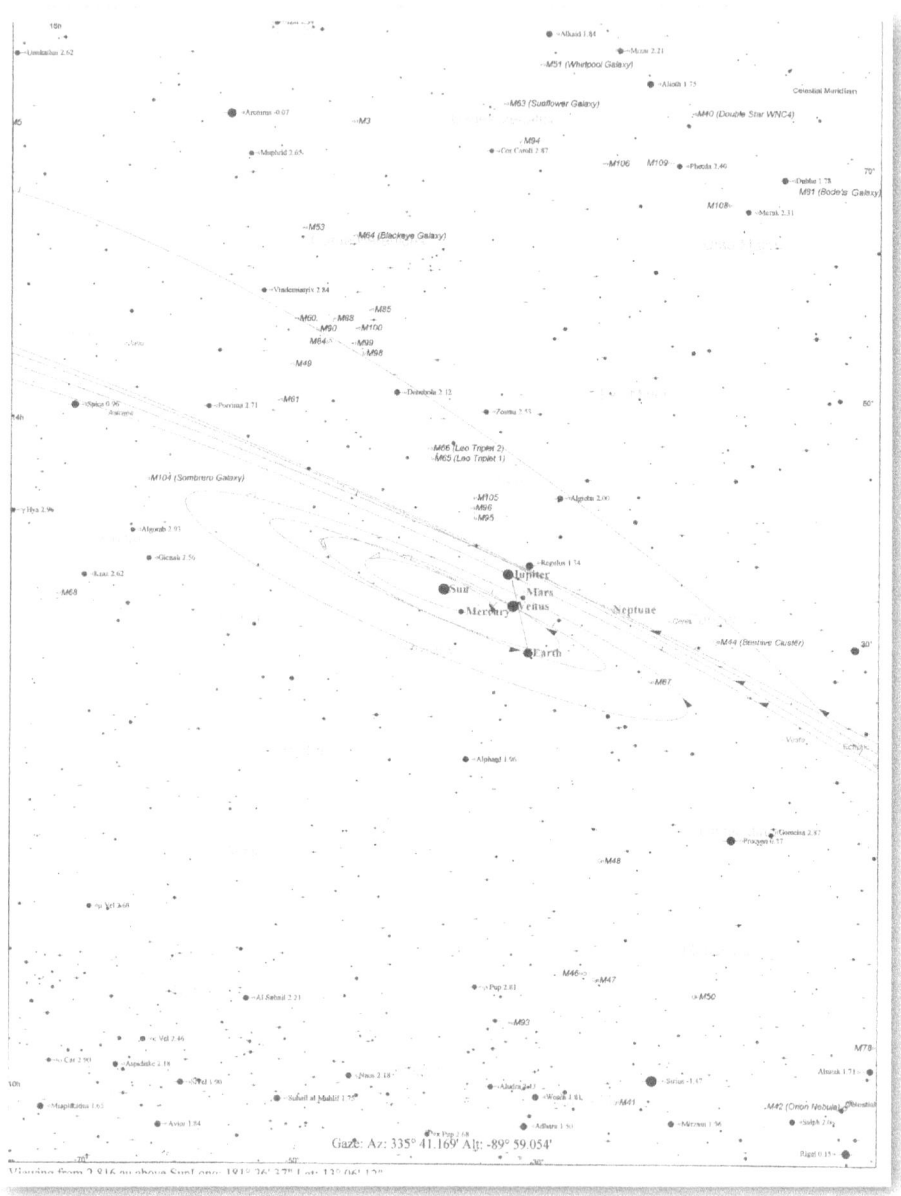

Future star, view from space. Image courtesy of Starry Night Astronomy™.

Hurriedly, I switched to the viewpoint of standing on earth at Jerusalem. *Yes!* It is a star, a future star! *Yes!* This is a star indeed, a second star, and that becomes undeniable. The time of the star is September 27, AD 2252, at 3:45 a.m. Three of these planets are the same that appeared at the birth of Jesus: Venus, Mars, and Jupiter in line. The future star looks similar to the star of Jesus, with the exception of Mercury; both stars are in the morning and in the east. *Wow! Wow! Wow!* The planet grouping continues very tight for about three days. It is one thing to know about the star of Jesus, as written in scripture, but it is another thing to find an unknown star, and so soon in the future. Later I found out the twenty-seventh is a Monday, and as before, I chose what I thought was the best view.

This star is occurring soon, very soon (235 years from now). My search for this star began in early 1996, and it is now 2017; it has been twenty-one years since I located this star and forty-two years since I first stood on the pier. To get the correct number of years before the future star appears, subtract the present year from the future year AD 2252.

Then I recalled what Jesus said in Acts 1:7–11: "[I]t is not for you to know the times or seasons." Wasn't Jesus talking to the apostles when he made that statement? Be assured the times and seasons have not been kept from *you*. Are both the Star in the East and this newfound future star co-inciding with events in scripture? Could this future star be a benchmark in time to show us something important? The birth star is for Jesus. What connection might Jesus have with this star? Each star seems to mark an arrival, and each must represent an event.

We now have about 235 years to go until the future star arrives. I am not Jewish, but this Jewish year is 5777, and their calendar stops in the year 6000. They have 223 years left on their calendar. A twelve-year difference before the future star appears that I cannot reconcile!

*– –|(*)

Future star, at Jerusalem, September 27, AD 2252, at 3:45 a.m.
Image courtesy of Starry Night Astronomy™.

What event does this star signify, and what will occur with the future star? What do the scriptures say?

> This is what Jesus himself said about this star: "I Jesus have sent mine angel to testify to you these things in the churches. I am the root and the offspring of David, the bright and morning-star" (Rev. 22:16).

The future bright and morning star has been an extraordinary find. What might be the big event? Will Jesus reign as king, as the scriptures tell? Is the planetary convergence the start of the kingdom age/millennium a star? To me, it is very clear when Jesus will come back and a new era of reign will begin. The age of grace may end, and deliverance through grace may be over. All this may happen when the future bright and morning star appears on September 27, AD 2252. Please let me add some perspective by using a verse written a couple thousand years ago. The next statement happens to be another from Jesus himself:

> And he that overcometh, and keepeth my works to the end, to him will I give power over the nations: And he shall rule them with a rod of iron; as the vessels of a potter shall they be broken to shivers: even as I received of my Father. And I will give him the morning-star. (Rev. 2:26–28)

After telling people about the upcoming future star, and they figure out when it arrives, I have gotten a consistent response: "I'll be dead by then." They refer to the future date in personal terms. This might be the greatest reason I wrote this to *you*. Perhaps you're saying to yourself, "I'll be dead by then!" Yes! In all probability, you will, and so will I. Please know there is hope that when you die, you can come back to life the same as Jesus did. Let me try to make that clear: *you* can only see the bright and morning star if you are alive! You say, "So? I'll still be dead by then." Yes, but…I told you before, only now I will be more emphatic: this is about *you*! The following ancient text may bring your *ah–ha* moment:

> But I would not have you to be ignorant, brethren, concerning them who are asleep, that ye sorrow not, even as others who have no hope. For if we believe that Jesus died and rose again, even so them also who sleep in Jesus will God bring with him. For this we say unto you by the word of the Lord, that we who are alive and remain to the coming of the Lord shall not precede them who are asleep. For the Lord himself will descend from heaven with a shout, with the voice of an archangel, and with the trumpet of God: and the dead in Christ shall rise first: Then we who are alive and remain shall be caught up together with them in the clouds, to meet the Lord in the air: and so shall we ever be with the Lord. Wherefore comfort one another with these words. (1 Thess. 4:13–18)

Please let me make this information as understandable as possible. Two major events with Jesus will happen in the future, according to scripture. Jesus will come back to get his church, and Jesus will arrive as king. The next verse stands as one of the more important verses I could have included. Why? Because when I tell some people when Jesus is coming back and when the future star will appear, they do not hesitate to say I am a liar—strong words. Then again, they do not have my insight into the future. To them, the scriptures say that nobody knows when Jesus will return, as in Saint Matthew: "But of that day and hour knoweth no man, no, not the angels of heaven, but my Father only" (Matt. 24:36).

The whole purpose of my writing has been for *you* to gain insight. What day or hour is being articulated—which? The future bright and morning star must be without a doubt not included in "that day or hour knoweth no man." You already observed the star, and you can see the star for yourself in an astronomy program.

To be as clear as possible, I feel the need to explain more about the terms "caught up" and "in the air" from the Thessalonians passage mentioned earlier. *When* "caught up" and "in the air" will happen may not be the utmost paramount piece of information. The important action necessitates that he will "gather his elect" and the statement that "I will raise him up at the last day." Who knows the day and hour when "he gathers his elect" and "I will raise him up at the last day"? No man knows, not even the angels in heaven.

Perhaps you may want a peek into the future at the time of the future star. Matthew in the twenty-fourth chapter paints a picture of a dark time that will be concluding. He continues describing the appearance of the sign of the Son of Man in heaven and the arrival of the Son of Man in the clouds of heaven. I feel certain that those appearances begin a new era referred to as the kingdom age or the millennial age.

As your guide, I say let the journey continue. We have discovered three *keys* and two benchmarks, bright and morning stars.

CHAPTER 3

STARS PAST - PAST
THE LAW
THE CONTRACT

To review, what we have established so far are the celestial quotation marks for the life of Jesus, consisting of a star and an eclipse. The discovery of a future star in AD 2252 that had been located with an inner-space point of view meant the time between the Star in the East and the future star had an interval of 2,251 years. I contemplated: could there be any significance to the 2,251-year interval? Would a fourth key be the 2,251-year time interval? As far as I knew, there were only two stars, and the time span did not mean a thing. Even so, I went searching for another star beyond the future star, using the 2,251-year interval to arrive at AD 4503. I found nothing.

*– –|(*)

Undaunted, I returned to the computer to look around in the past/time before the Star in the East using the 2,251-year interval. The obvious next step would be a trip back in time to 2251 BC, though I felt I might be wasting my time. My expectations of finding anything were nonexistent. Because, earlier, I did not find a future-future star at AD 4503 by using the 2,251-year interval, why would I expect to find anything in the past/time using the same interval of time? Wouldn't historians have recorded any stars, if they had the knowledge? With the inner-space viewpoint, pattern recognition, and the 2,251-year interval, I watched the planets slowly circling the sun to see if a pattern of planets lined up. I hit the jackpot—more than I could have possibly imagined!

There were two stars twenty-two days apart, an unforeseen discovery. A surprising gift, as I thought I knew the past. At first, I got these two stars confused. Again, I went to Jerusalem to verify that both convergences were indeed stars. Yes, each set of planets is a star. After I found them, I had no idea why they existed; it baffled me. It took me a while, but this agreed with what I came to conclude by using scripture. One is another morning star in Jerusalem and in the east—a new beginning and a third benchmark in time. This is the *third* morning star I found, and by now, the morning-star patterns are recognizably similar. The other star, in the evening, has been the only evening star I found.

The third morning star consists of a conjunction of Venus, Jupiter, and Saturn. With the discovery of this star, the 2,251-year interval gained a lot of importance; it would allow for more exploration, and a timeline has begun to form. We have already learned that morning stars mark major events. Why wasn't the star recorded in history, and what significant event occurred? I believe the event was recorded, just not the star—but *why?* What follows is the record in scripture when God gave the Ten Commandments on January 5, 2250 BC, at 6:30 a.m. "And all the people saw the thunderings and the lightnings, and the noise of the trumpet, and the mountain smoking: and when the people saw it, they removed, and stood afar off" (Exod. 20:18).

The arrival of the Ten Commandments star began a new era of *law*.

– –– –| (*)

Ten Commandments star, at Jerusalem, January 5, 2250 BC, at
6:30 a.m. Image courtesy of Starry Night Astronomy™.

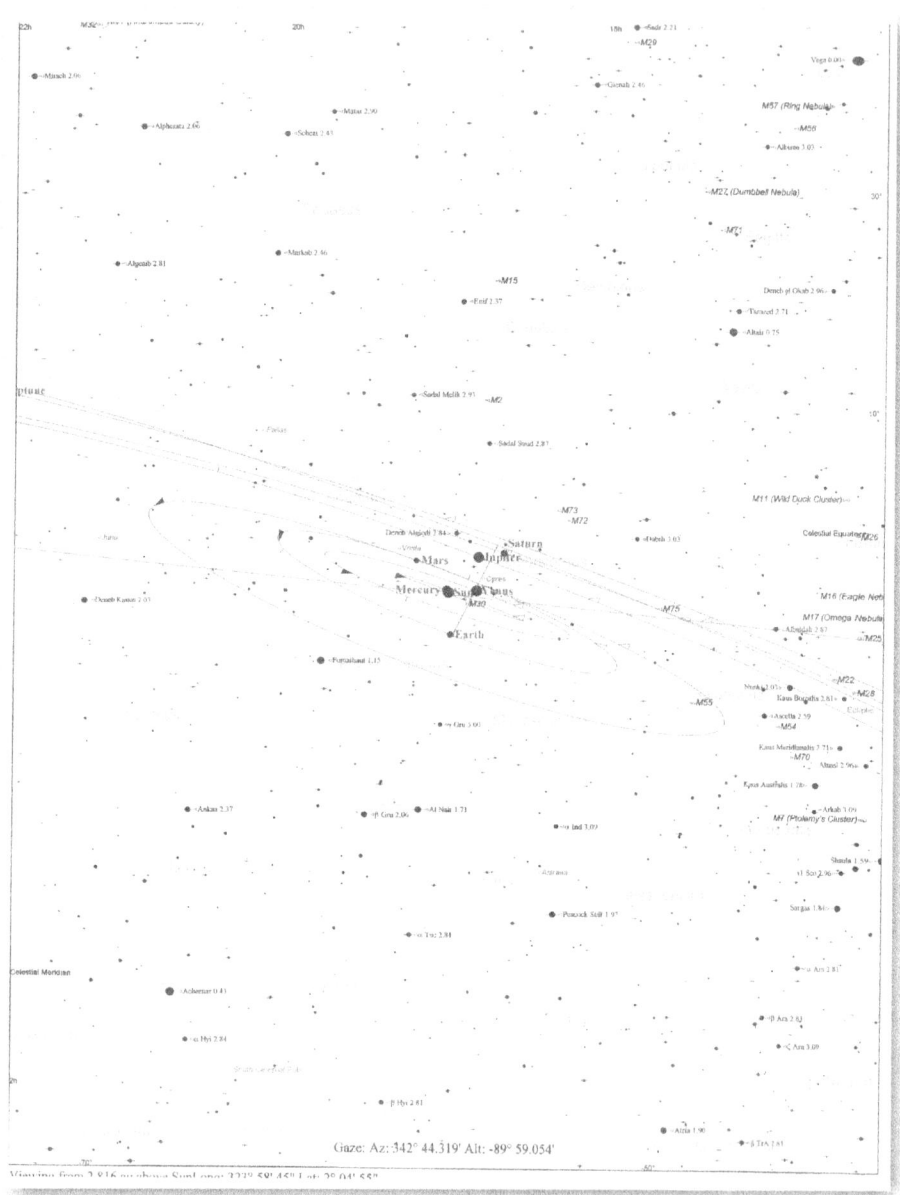

Ten Commandments star, view from space. Image courtesy of Starry Night Astronomy™.

The Jewish calendar records the counting of the Omer, forty-nine days or seven weeks, which begins on the second day of Passover. The counting of the Omer begins the time between Passover and the Ten Commandments or Shavuot. Therefore, the first Passover began fifty days before the Ten Commandments' star. The convergence lasts for about three days, so the exact day would be uncertain. Passover is significant because on that day, the Jewish people departed Egypt.

The evening star occurs about twenty-two days earlier than the third morning star. It is a conjunction of Jupiter, Saturn, and Mars in the southwest that may have been at the time of the exodus. The evening star appeared on December 13, 2251 BC, at 5:00 p.m., plus or minus a day. My best guess is that the aligned planets helped part the Red Sea during the exodus. The evening star may mark the crossing of the Red Sea, and it does not appear to be part of the timeline. It is possible there are more evening stars; this one was uncovered only because it was so close to the Ten Commandments star.

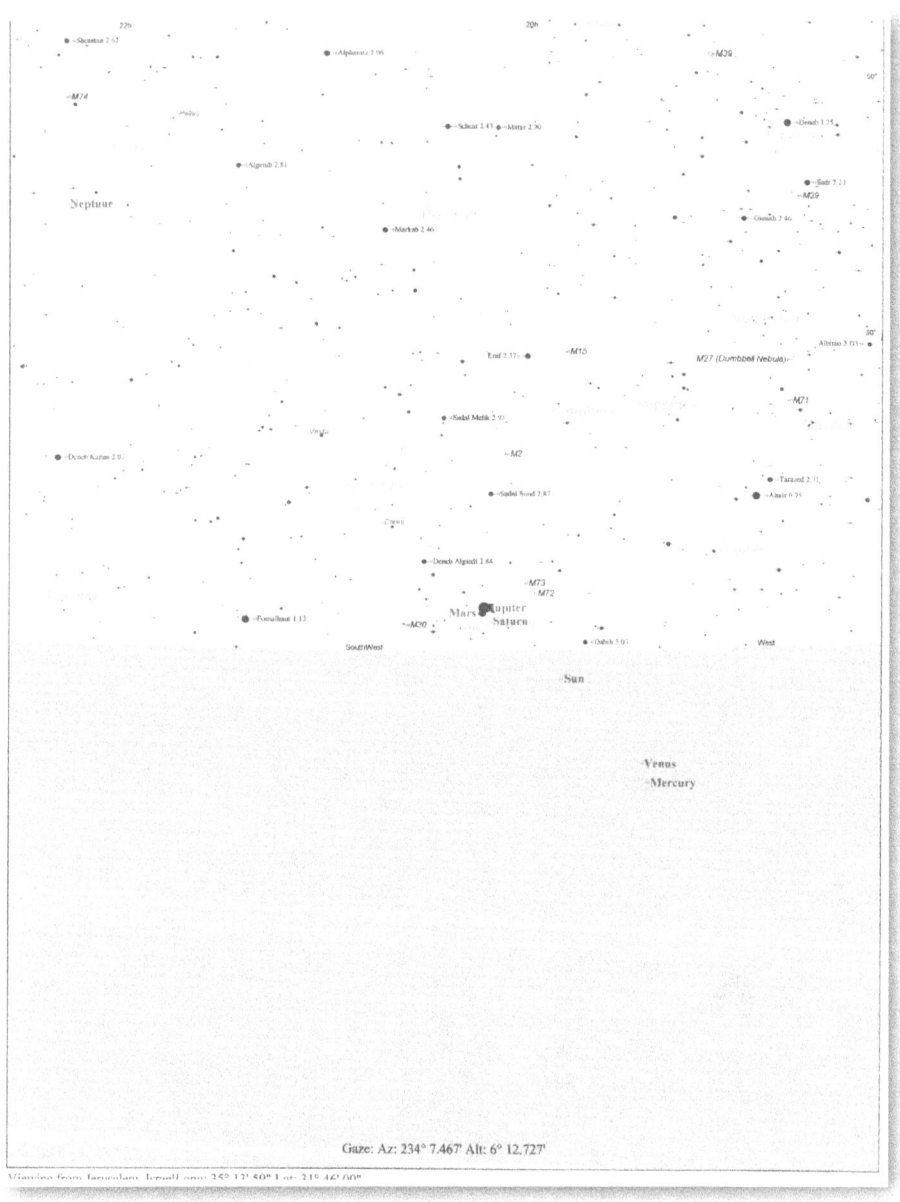

Evening star, at Jerusalem, December 13, 2251 BC, at 5:00 p.m.
Image courtesy of Starry Night Astronomy™.

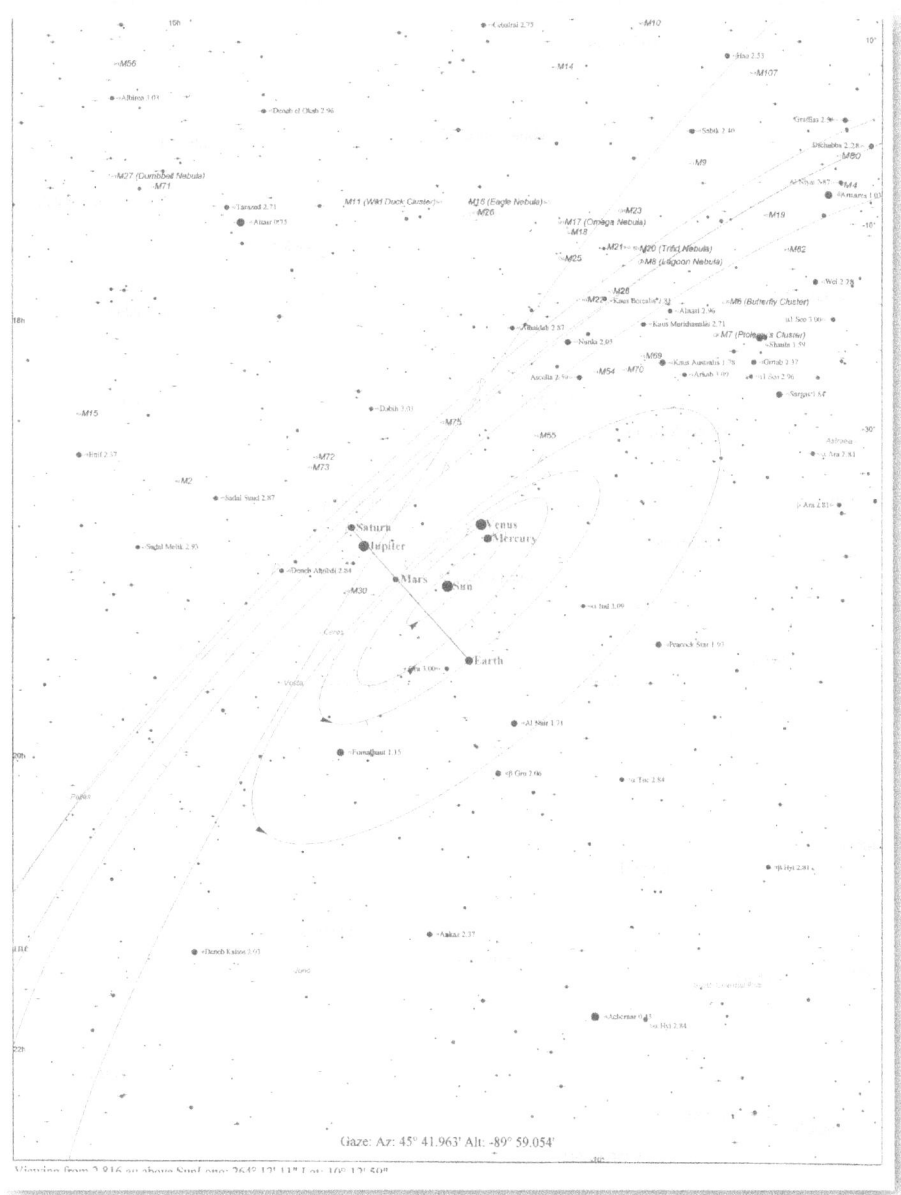

Evening star, view from space. Image courtesy of Starry Night Astronomy™.

More time travel into the past is coming, and we have four keys to try out. We're not done with the tour; let's keep going. We have more to find out about the past and the future. The present is still ticking…and each passing tick vanishes into the past.

I am starting to realize that morning stars mark important events, new beginnings, and benchmarks that form a timeline. The next step would be to go back in time, using the same interval of 2,251 years, to reach 4500 BC and search around using the keys. When a matching pattern came into view, I became excited to locate it, of course, and deduced very fast that this would be another star when seen in Jerusalem. The star comprised four planets, and I recognized it as a major star.

Four planets—Venus, Saturn, Mars, and Mercury—converged on October 5, 4497 BC, at 4:45 a.m., a bright and morning star far back in time. Again, the date picked was what I thought as the best view. This is my *fourth* discovery of a morning star viewed from Jerusalem and in the east. It stood as another benchmark in time, and my searching led to try determining why it should have been there. The creation of Adam and Eve seems plausible, doesn't it? It marked the beginning of the era of sin. If the program went further back in time, I can't remember, but even so, I didn't find any more stars in the past. It's hard to conceive how God arranged the planets into stars and the events to coincide at the same time. Are you beginning to get the scale of this timepiece? I do not know where the Garden of Eden existed, but Jerusalem happens to be a good viewing point to see the star.

A very long time ago, there lived a rich man called Job, blameless and upright. He was the greatest man in the East, and he lost everything—his children, herds, health, and status—and he questioned God: Why? God answered Job out of a cloud with a series of questions about the creation of the universe, and one of the relevant questions to Job for us must be: where was he…? "When the morning stars sang together, and all the sons of God shouted for joy?" (Job 38:7). Job declared he did not have understanding. God blessed Job with twice as much as he had before, because during his sufferings Job remained blameless and upright and did not blame God.

∞*– –*– –*– –|(*)

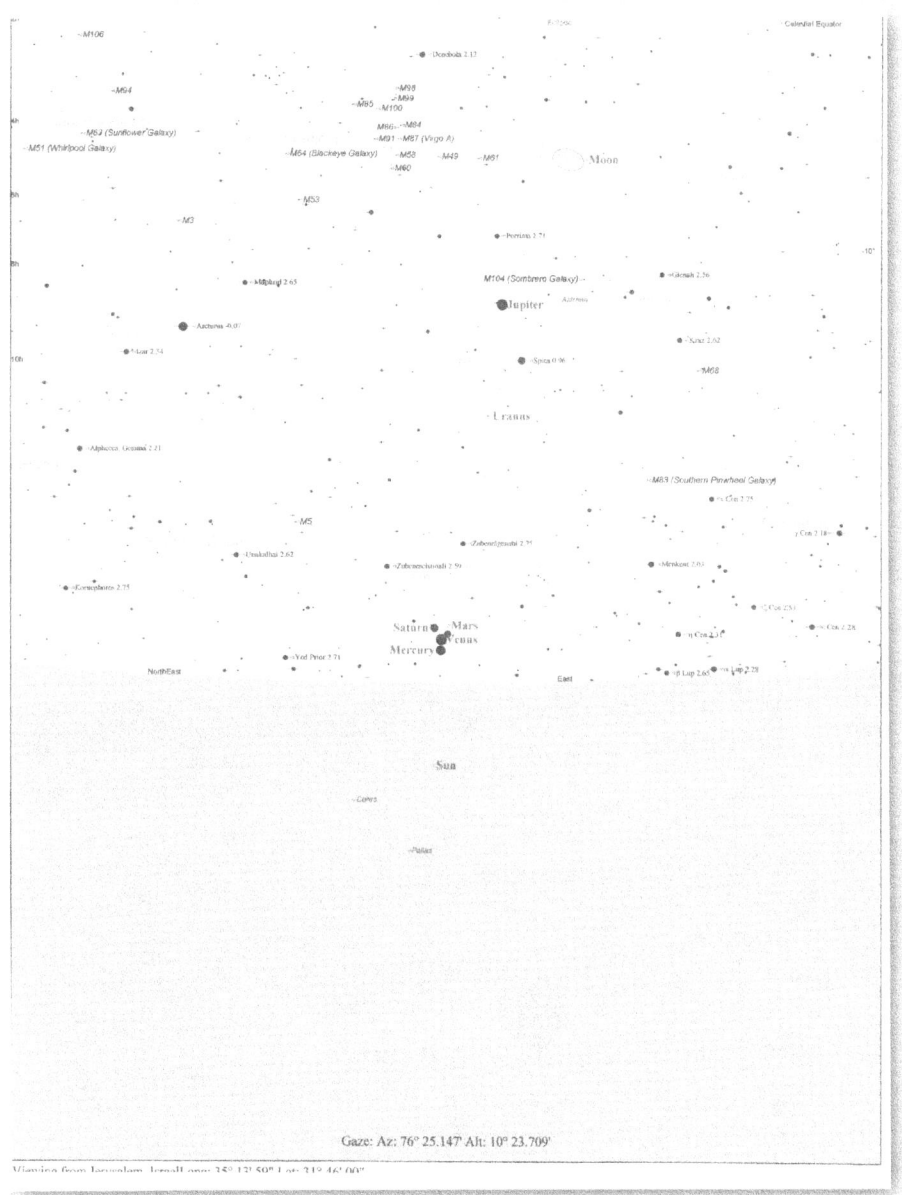

Adam and Eve star, at Jerusalem, October 5, 4497 BC, at 4:45
a.m. Image courtesy of Starry Night Astronomy™.

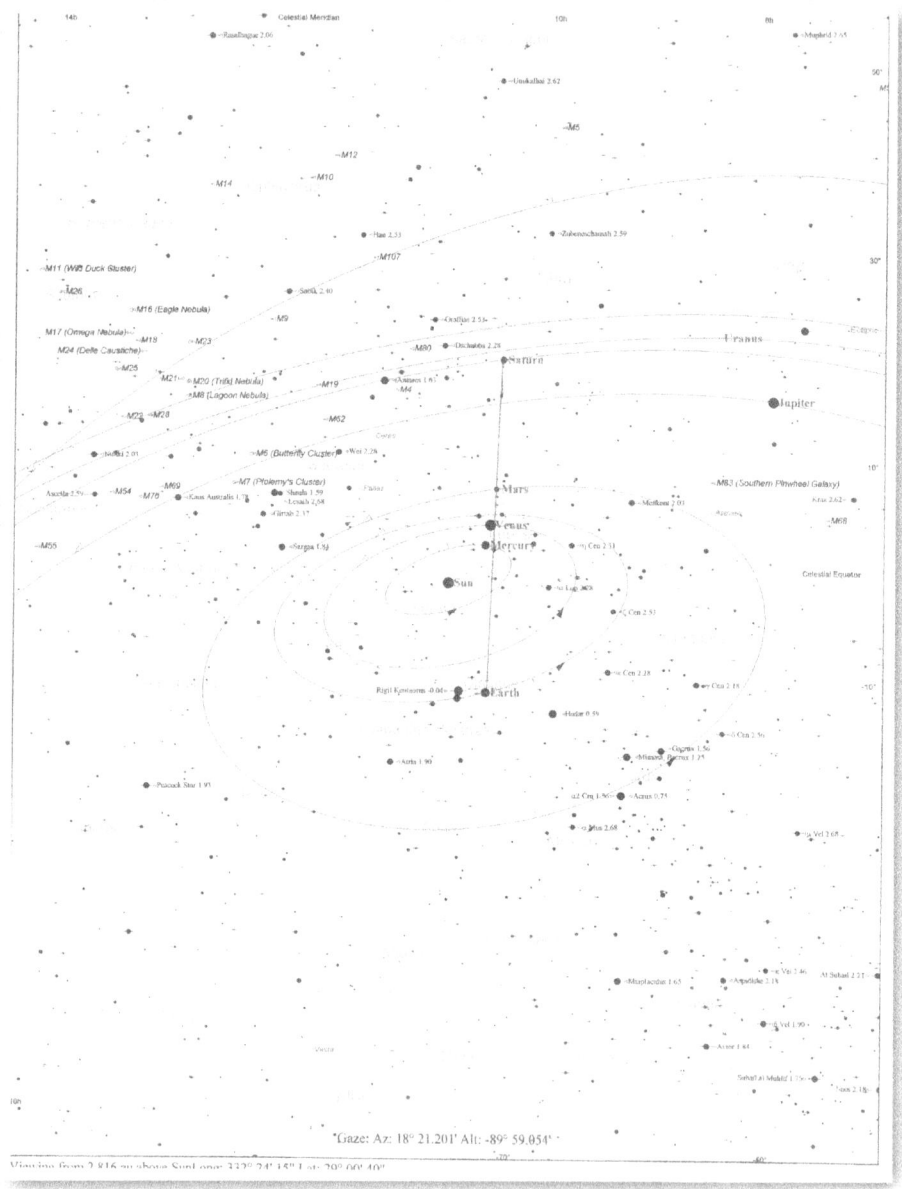

Adam and Eve star, view from space. Image courtesy of Starry Night Astronomy™.

Now, I know anthropologists and archeologists spend their whole lives in esteemed and valued study, and I am optimistic that all the stars can be aids in their work. I believed something extraordinary happened when the star of Adam and Eve appeared. They lived in a garden and in the middle of the garden were the tree of life and the tree of the knowledge of good and evil. My best guess has been that God made a contract with Adam and Eve. A contract consists of an *offer*, to eat of any tree in the garden, an *acceptance* of the conditions, you must not eat of the tree of the knowledge of good and evil for you will surely die, and a *consideration* in which each party to the contract must receive something of value. Adam and Eve received "the promise to live eternally sinless." What did God receive of value? "I will be their God, and they shall be my people." This contract Adam and Eve broke, as we know, by not living a sinless life. They defaulted on the contract, so that everlasting life could no longer be valid. Death, the only option, is the result. God's part of the contract, the consideration, became null and void as well because of Adam and Eve's sin. For God to receive his consideration and redeem his people, "I will be their God, and they shall be my people," the penalty of death, in order to be paid, would require a sinless death.

Jesus, who was without sin and the other party to the original contract, cried out, while dying on the cross, "It is finished." What was finished? He had to give his own life to honor the conditions of the original contract to Adam and Eve and us, which is "the promise of living eternally sinless." Only Jesus, by a sinless death, can redeem them and us, with "the promise of living forever without sin." Adam and Eve did not live sinlessly, and neither can we. Jesus redeemed the contract, and only Jesus can offer grace to restore God's consideration. "I will be their God, and they shall be my people."

Jesus is the Christ, a Jew, and returning as king. Right now, everybody who would be willing can receive grace from Jesus by simply asking for forgiveness of sin. He will do it. This is grace. Then, when you ask for grace, you become a party to a contract that has an *offer*, an *acceptance* of the conditions, and a *consideration* in which each party of the contract must receive something of value. *You* receive forgiveness of sin by *grace* and the promise of living an everlasting, sinless life. What does Jesus receive of value? "I will be their God, and they shall be my people."

Revelation 22:17 says, "And the Spirit and the bride say, Come. And let him that heareth, say, Come. And let him that is thirsty, come: And whoever will, let him take the water of life freely."

∞*– –*– –*– –|(*)

Have you considered that using the Star of the East and the interval/time span of 2,251 years has enabled us to locate three more stars? Let's review our travels so far. I used the infinity symbol to represent the time before Adam and Eve's star. In that symbolic time, it could be possible that more stars exist, but I could only go so far with the program. Beginning with the Adam and Eve star, we next have 2,247 years until the Ten Commandments star; we then have 2,249 years until the star of Jesus, and after that, we have 2,251 years until the arrival of the future star in AD 2252. Where are *you* in time? Subtract your current year (this year is AD 2017) from the future year AD 2252, and there are presently 235 years until the arrival of the future star. Now *you* have an idea of your place in time. I believe that the proximity of the future star puts us close to the end of the age—the age of *grace*. The Bible has many verses on the end times, which we can be near, if we are not there already.

As your guide, I want you to understand that we live in an extraordinary timepiece, and the time is ticking.

CHAPTER 4

MYSTERY, STAR
TIMELINE
DAYS OF GOD
MAGNETIC REVERSAL

Not long ago, by chance, I had an unexpected find of a future-future star that will occur on May 10, AD 4999, at 4:00 a.m., with four planets: Venus, Saturn, Jupiter, and Mercury. This happens to be my *fifth* morning-star discovery viewed from Jerusalem and in the east. As this star has four planets in its appearance, and that fact must make this star a special new beginning, a very important benchmark. The only other stars with four planets are the Adam and Eve star and the star of Jesus, and now this star. What it is for, I don't have a clue; it remains a mystery. Finding this star involved luck, because it does not follow the pattern of the time span I discovered. Unlike the other stars, this star has a time span of 495 additional years. Perhaps this star could begin God's first workday. In the condensed version of the timeline, the ellipsis is representative of the time beyond this future-future star. Having reached the limits of my program, this was the last star I found. Are there more stars?

$$\infty^*\!-\!-^*\!-\!-^*\!-\!-\mid(^*)\!-\!+((^*))\ldots$$

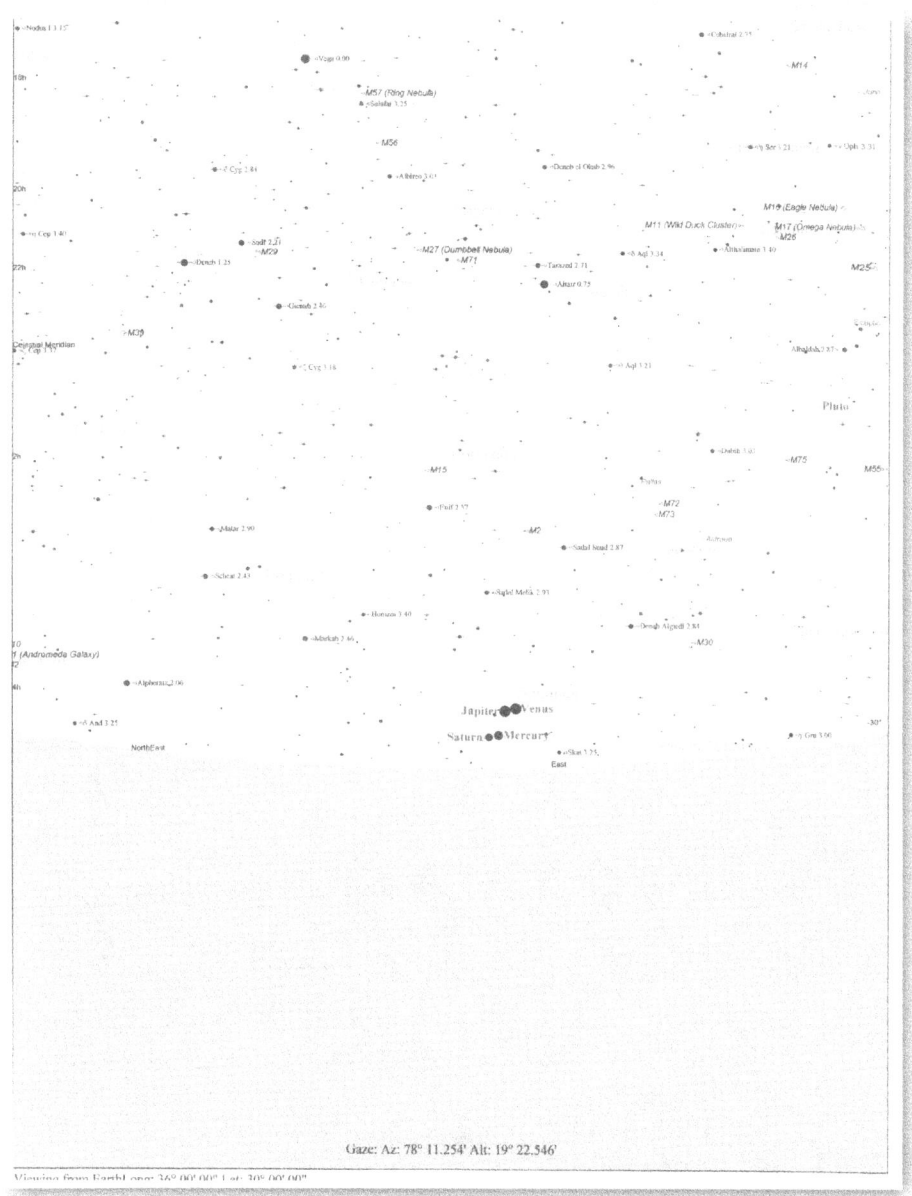

Future-future star, at Jerusalem, May 10, AD 4999, at 4:00
a.m. Image courtesy of Starry Night Astronomy™.

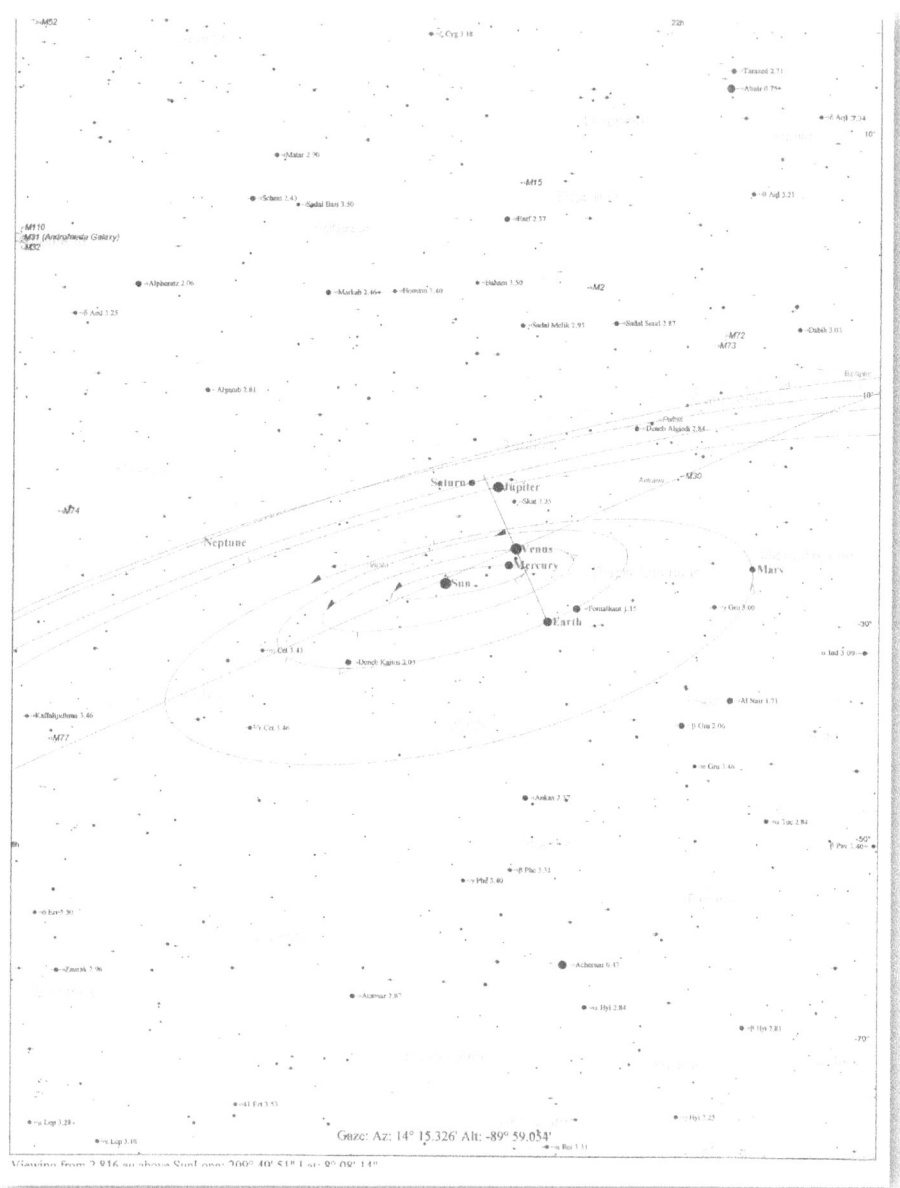

Future-future star, view from space. Image courtesy of Starry Night Astronomy™.

Previously, God answered Job and questioned him: "Where wast thou when I laid the foundations of the earth? declare, if thou hast understanding" (Job 38:4).

You have seen how the stars form a timepiece and are part of a time-line. Complicated, yes. The timepiece spans from 4497 BC to AD 4999, a total of 9,496 years, and uses all five visible planets. And time is still ticking. Now *you* know where *you* are in relation to the eons of time. I've already shown that the eras after the stars have meanings. Adam and Eve's star began the era of *sin* and the Ten Commandments' star the era of *law*. Jesus's birth brought us the era of *grace*. The future star begins the *kingdom age*, and the future-future star? A *mystery*!

My guide of the stars should be complete; however, it's not over, and we're not done yet. There's more! A timeline of benchmarks offers other insights when using the scriptures. Scripture tells us the length of the Lord's day: "But, beloved, be not ignorant of this one thing, that one day is with the Lord as a thousand years, and a thousand years as one day" (2 Pet. 3:8).

The middle time distance between the stars is about 1,125 years—a little over a thousand years—so for God there are two days between the stars. We know that God's seventh day, a day of rest, is after the star of Adam and Eve as told in scripture: "And on the seventh day God ended his work which he had made; and he rested on the seventh day from all his work which he had made" (Gen. 2:2).

If we count God's days after the star of Adam and Eve, we are near the end of God's fifth day. In scripture, numbers can have a meaning. The number five also has the meaning of grace; this is the age of grace, the fifth day.

Six is the number of man, and this is man's sixth day when using God's standard. As I understand it, man's rule consists of six days; if that is the case, his days to rule are almost over.

Numbers in scripture can represent abstract concepts. *One* represents unity—as one God. *Two* alludes to separation and union—examples are man and woman, and two witnesses. *Three* suggests divine perfection—as in Father, Son, and Holy Spirit. *Four* relates to creation—as in four seasons. *Five* corresponds to grace. *Six* refers to mankind—such as he shall labor six days in a week. *Seven* refers to perfection—as seven days make a week. *Ten* reveals law and testimony. *Twelve* asserts government. *Forty* refers to being tested. These and many more numbers have significantly shown that their repeated use in scripture indicates that concepts are associated with numbers.

The last bit of information to glean from the timeline is about the earth. Have you seen a map and wondered what all those lines are in the ocean bottom or why they are there? Studies of the sea floor have led to the conclusion that magma is pushing up and has hardened to form ridges. These ridges recorded the polarity that existed as they hardened. The ridges are in rows of alternating polarity, and each row requires about five thousand years to make. The conclusion is that the earth reverses its magnetic field approximately every five thousand years. In the future, when the reversal happens, the North Pole will become the South Pole and vice versa. After the future reversal, when the next five thousand years are finished, the poles return to the same direction as now. Our solar system takes ten thousand years to make one magnetic rotation, and judging by the number of ridges arranged on the ocean floor, the earth has been through many rotations.

When Adam and Eve walked the earth, the magnetic fields were opposite than they are now. The Bible genealogy documents the number of years to Noah's flood. Noah's flood came 1,656 years after Adam and Eve's star. When will the next reversal happen? Assuming Noah's flood was the last magnetic reversal, we can figure out when the next magnetic reversal occurs. The years after the flood to the future star add up to 5,091 years, and 5,000 years is our target until the next reversal. We can conclude the next magnetic reversal is before and near the time of the future star. I suspect that volcanoes, earthquakes, and solar flares may increase, and perhaps other phenomena. This week there were large earthquakes in

Japan and Ecuador. Isaiah 24:20 says, "The earth shall reel to and fro like a drunkard, and shall be removed like a cottage; and the transgression of it shall be heavy upon it; and it shall fall, and not rise again." Could Isaiah be referencing the upcoming magnetic reversal?

Searching the program midway between the stars, I found nothing, but I continued to search the scriptures for more astronomy-related verses like the "three days of darkness" in Exodus 10:22 and "the sun to go down at noon" in Amos 8:9. My search for the three days of darkness turned out to be very extensive and time-consuming. I searched through years and years in the inner space view for Venus and Mercury to block the Sun from reaching Egypt for three days. I found nothing.

If you're interested in the time between now and the future star, the Bible can tell you a lot. I hope I have encouraged your interest. Now you've been presented with the big portrait, and it is all for *you*!

Perhaps you're noticing, or maybe it didn't occur to you, that I have known this information a long time. Why hasn't this information come out earlier? I tried. I did my best, but the knowledge never got any traction. As I write this, it may still not. The only difference this time involves having tried to explain with more background details. Having sent or passed on this information to many places—and the list is extensive—I do not re-member ever receiving a reply or so much as an acknowledgment. This has led me to believe that people are not able to spend time on something not that relatable, and they do not understand. Sometimes at Christmas, news-papers would publish an article trying to explain the star of Bethlehem, and I gave up trying to correct them, because the next year they repeated the same old article. One year, my wife and I went to the local college for a Christmas program at the planetarium. They presented a star for Bethlehem that consisted of two planets high up in the sky. They had a later/wrong date. Disappointed, I spoke up in the crowded auditorium and asked the host of the planetarium if she could change the date to a different date on the machine. I knew the date, and I sure wanted to see that star! The host said, "No, it takes a long time to set up the machine." I do not think she appreciated my interruption. After the program, I went home, got a copy of the star info, returned, and gave it to the host. That experience was years ago, and to this day I have yet to see the stars in a planetarium. Someday I expect to see the stars in a planetarium, because they can provide the realism of actually being there, by showing the atmospheric effect caused by air molecules refracting the light to make the stars larger.

We are nearing the end of our journey, and as your guide, I want to thank you for staying with me. I trust you have or can get a new guide, a preacher who understands the Word of God.

In keeping my promise that this knowledge concerns *you*, and with the timepiece ticking now, how will *you* respond? What is your choice? What is your destiny? Job in 38:7 tells us about "when the morning stars sang together," but now in your mind's eye, you can see the star, and it is real to you. That same verse goes on to say, "And all the sons of God shouted for joy." I hope *you* shout for joy as well.

Look at the timeline drawing of stars, and you will see a progression of stars, benchmarks, to events that happened during the past and will happen in the future. The layout is as accurate as possible so you can get an overall visual representation of topics written in the book.

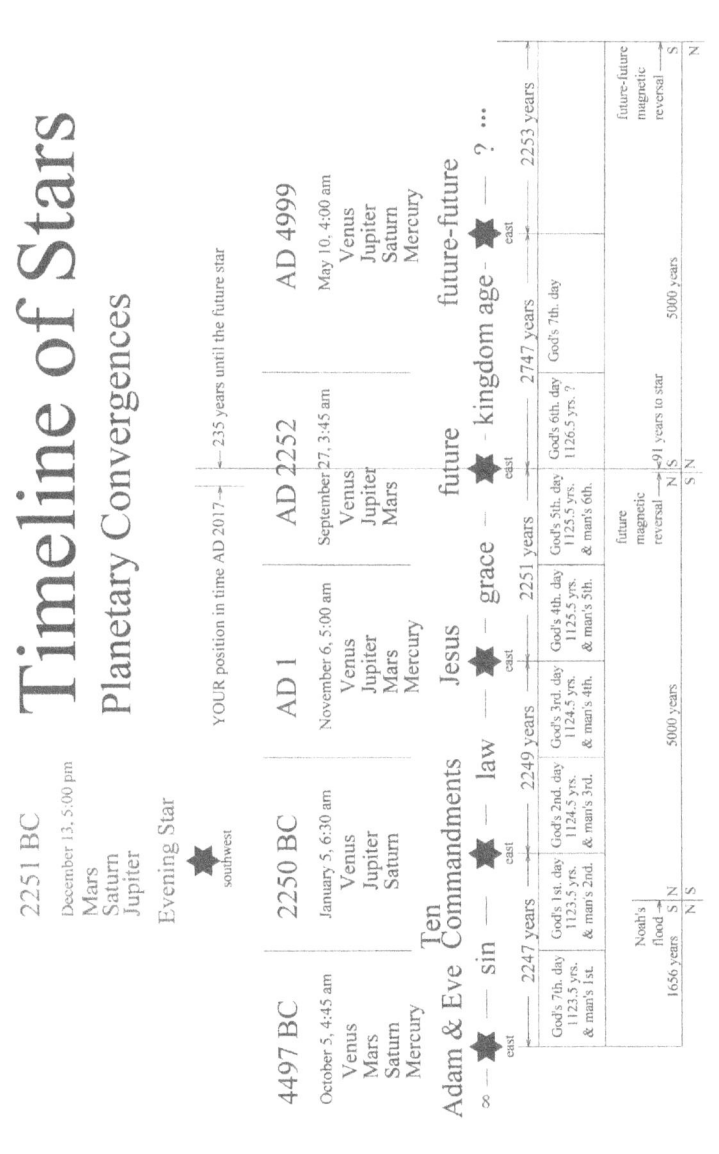

Timeline drawing.

These stars and the events connected to them are predestined to happen, and there are at least two more stars in the future. No power on Earth can stop the stars or the events from occurring. God has predestined their occurrence. Isaiah has some verses about God. Isaiah 55:7–9 says:

> Let the wicked forsake his way, and the unrighteous man his thoughts: and let him return to the LORD, and he will have mercy upon him; and to our God, for he will abundantly pardon. For my thoughts are not your thoughts, neither are your ways my ways, saith the LORD. For as the heavens are higher than the earth, so are my ways higher than your ways, and my thoughts than your thoughts.

A word of caution: If you're at a point where you're saying to yourself, *I need to tell this person or that person about the stars,* those persons have nothing relatable for them. The information is new and unknown to them. I hope you do better than I did. In all the years I have shared information about the stars, here are the responses I have come to expect: "That's nice," said in a dismissive manner; "You're a liar"; no response; "No, it's not, because…"

You might say, "But I should tell so-and-so." Do it. Remember, though, that it is hard to change lifelong beliefs when people do not connect to the subject.

Another word of caution: To those who persecute others, perhaps you might consider rethinking your ideas. What is your destiny going to be?

> Luke 10:24 says, "For I tell you, that many prophets and kings have desired to see the things which ye see, and have not seen them; and to hear the things which ye hear, and have not heard them."

Next are four renditions from the early astronomy program that I first saw probably in 1996. The four morning stars are in the order I found them. I wanted you to see replicas of what I first observed.

Star in the East, at Jerusalem, October 22, AD 1, at 3:50 a.m.

The first star I located was the Star in the East when Jesus was born. This drawing is important because you are seeing what I saw, minus the orbit lines. Please notice that the date is a little different from that on the timeline drawing. Notice, also, that the timeline drawing has four planets, and this drawing has three.

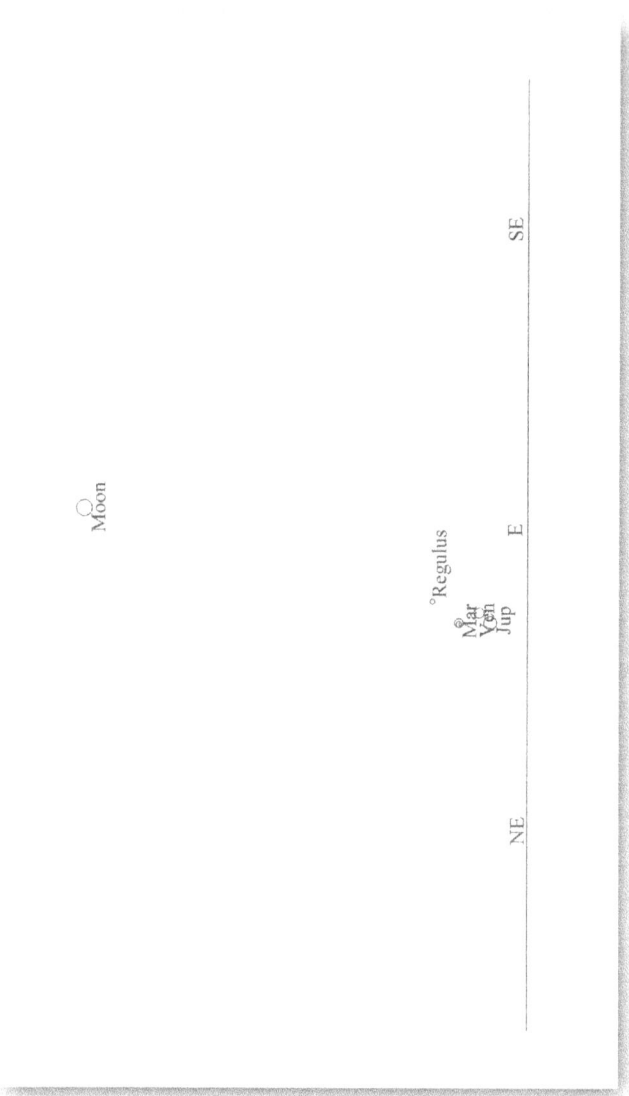

Future star, at Jerusalem, September 27, AD 2252, at 4:00 a.m.

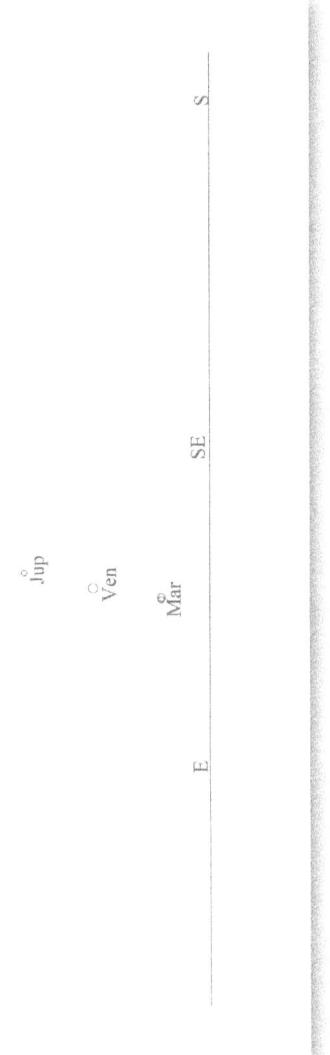

Ten Commandments star, at Jerusalem, December 13, 2251 BC, at 4:45 a.m.

The third drawing is of the Ten Commandments star. The date does not match the timeline drawing, the grouping of planets should be together, and they are not together.

Adam and Eve star, at Jerusalem, August 5, 4502 BC, at 8:30 a.m.

The drawing of the Adam and Eve star was the fifth star found, if you include the evening star in the count. It has many inaccuracies compared to drawings made by programs that are more accurate. The run times with my early computer program were slow. After starting the program, I would walk up to the marina shower house, shower, and still wait for the computer as the astronomy program progressed back in time, even after I got back to the boat.

Although these early drawings are not absolutely correct, the views and time spans were enough for me to detect a pattern. This section of the early astronomy program is completed.

∞*– –*– –*– –|(*)– +((*))...

Epilogue

When I was forty years old, I made a decision, a promise to myself, to read the Christian Bible cover to cover. I was not going to spend the rest of my life without having read the Bible. Period!

When I began to read the Bible, it was during my sailing era—a time of yacht-club luaus, parties, cruising, and racing. I loved that era, and my family remembers it well. It was fun, lots of fun. I began at Genesis, and inside myself, I liked the way I did things. I soon divorced; "I liked the way I did things" didn't work out so great after all.

In reading the Bible, I felt like it plodded along. For me it seemed at times as though I had just read the same thing in the last chapter. The promise to myself was the only thing that kept me going sometimes. I am not trying to discourage you—quite the opposite. I want you to be aware of my pitfalls so you do not feel as I did.

God makes many promises to you in the Bible, and one of those is that *his* words will not return void. First, if you are over forty, it is never too late to start reading the Bible. It will make you wiser. Second, if you are under forty, now is the time to discover the promises of God. My suggestion is to use study guides. I think my way was the hard way, but even so, *his* word did not return to *him* empty-handed. If I had known better, I probably would have started in the New Testament with the four gospels. Matthew, Mark,

Luke, and John are four writers with four points of view on the same years. As a first-time reader, you may not notice slight differences, and it may seem to you that you just read the same thing.

If I had not studied the whole Christian Bible, I might have missed the connection to the stars and many other things. I am still studying the Bible and probably always will.

You…

BIBLIOGRAPHY

Astronomical images courtesy of Starry Night, a Simulation Curriculum Company, Minneapolis, MN, all rights reserved, www.starrynight.com.

Bible Verses in Order of Appearance

Introduction
Luke 10:24

Chapter 1
Matthew 2:2
Luke 9:50

Chapter 2
1 John 4:9
Matthew 27:45
John 20:29
Acts 1:7–11
Revelation 22:16
Revelation 2:26–28
1 Thessalonians 4:13–18
Matthew 24:36

Chapter 3
Exodus 20:18
Job 38:7
Revelation 22:17

Chapter 4

Job 38:4
2 Peter 3:8
Genesis 2:2
Isaiah 24:20
Isaiah 55:7–9
Luke 10:24

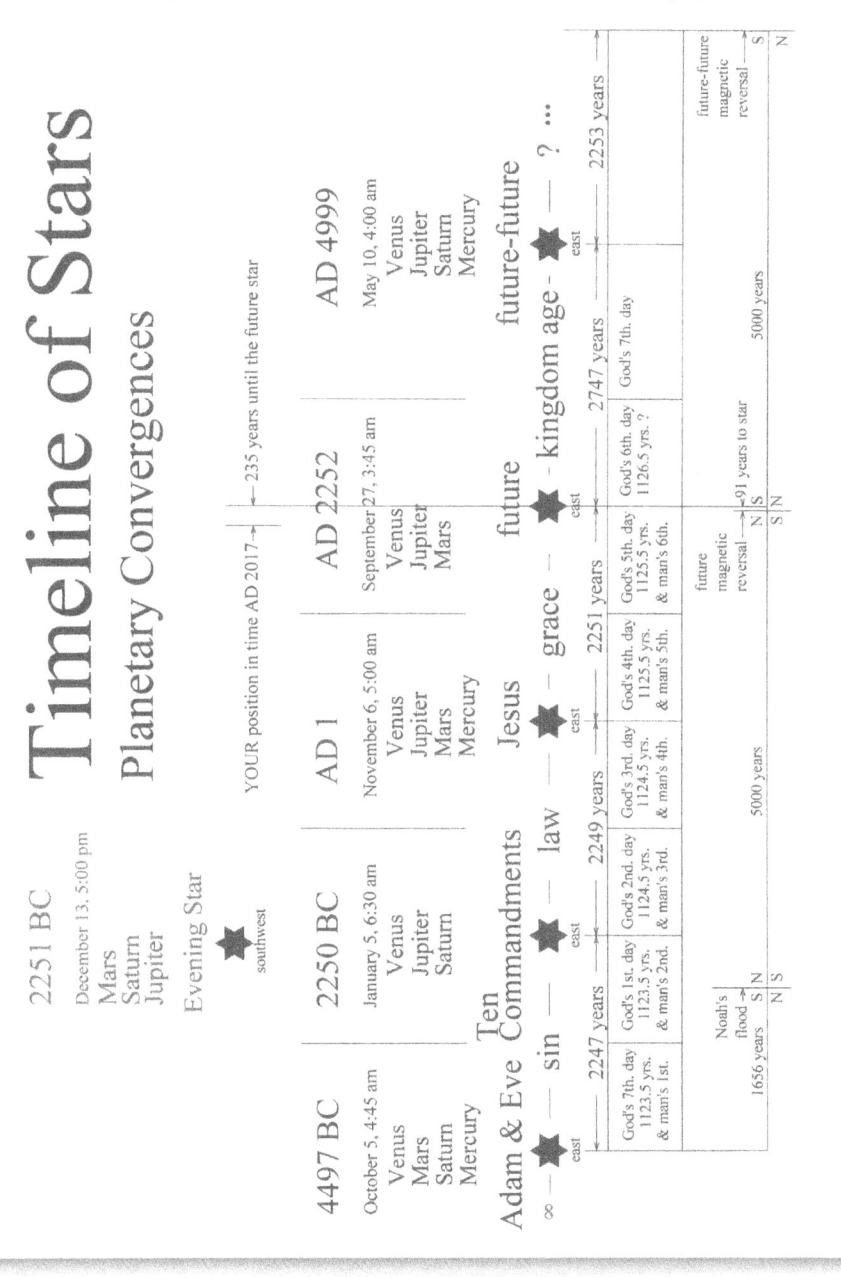

Timeline drawing.

About the Author

When he's not studying the stars or scripture, Joe Adcock can be found sailing or spending time with his partner and wife, Barbara. He makes his home in Southern California. *Tic, Tic, Tic, Tic, Tic* is his first book.

www.ingramcontent.com/pod-product-compliance
Lightning Source LLC
Chambersburg PA
CBHW071757170526
45167CB00003B/1065